How Grandma Got Better After *The Stroke!*

By

Valerie Clary Muronda

MASAKA PUBLISHING
MEDIA HOUSE

ISBN 978-1-965398-24-1
© Valerie Clary Muronda

Illustrations by Arjunnaja Pratama

My grand mom is my best friend! Since I can remember, my grandmom has been a big part of my life. She's been there when I get home from school, there for birthdays holidays, and all the important events in my life.

Today, mom has to work late, and dad is away for work stuff.

"Grandmom will be here when I get home from school! Yay! Perhaps we will make my favorite-chocolate chip cookies!"

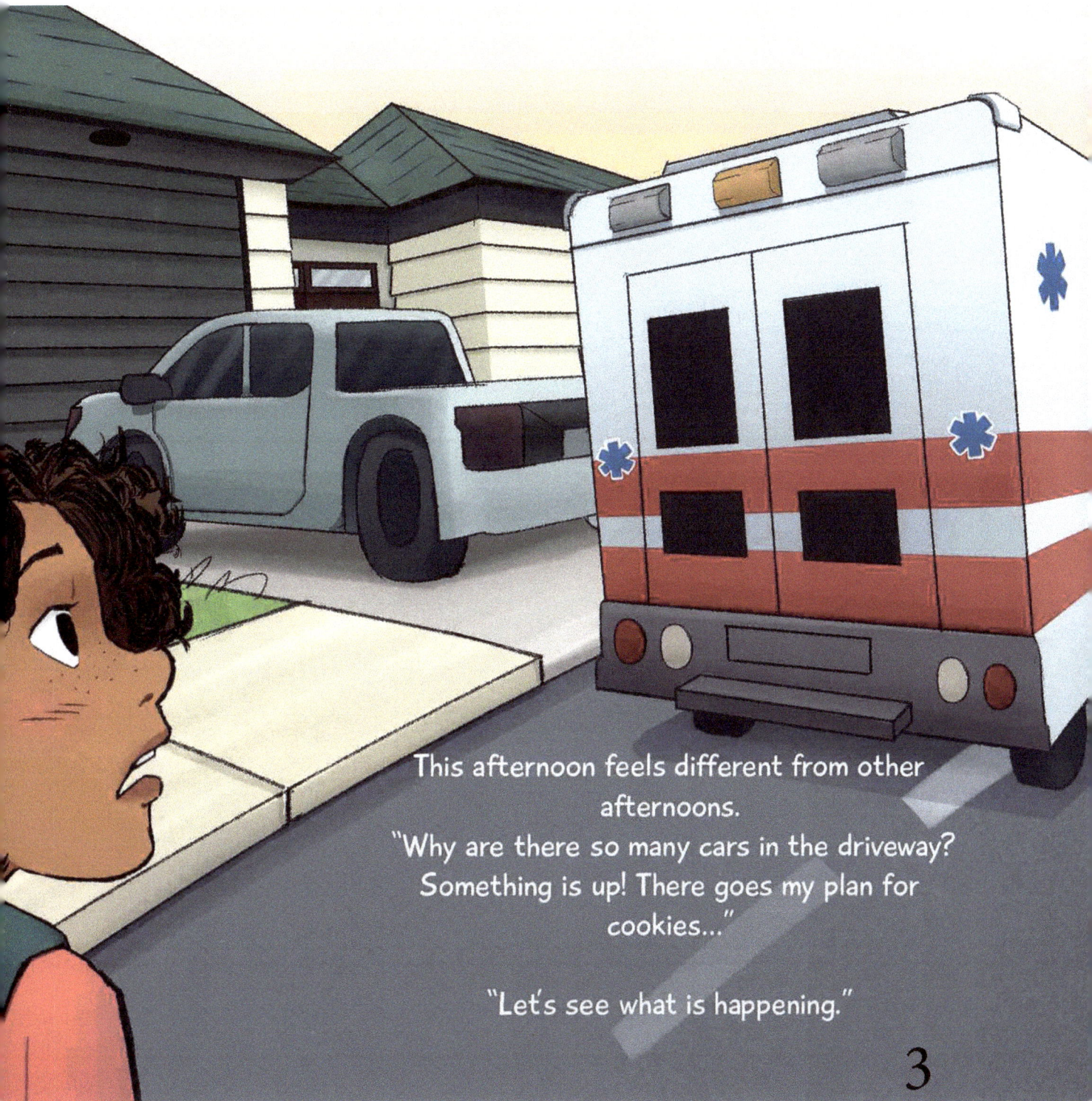

This afternoon feels different from other afternoons.
"Why are there so many cars in the driveway? Something is up! There goes my plan for cookies..."

"Let's see what is happening."

"Honey, I have to talk to you about something,"
Mom says in a concerned voice.

"Wait a minute, she's supposed to be working late..."

Mom continues, "It's grandmom, She go sick and she is in the hospital?"

4

"Oh no! No,no,no,no! What happened?! I asked?"

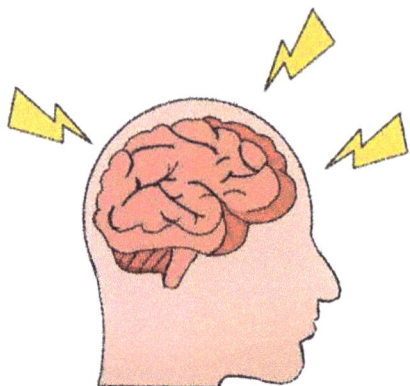

"Well honey, she had something called a "stroke" it is something that happens in a person's brain, and sometimes, they can have difficulty talking, walking and doing the things they were used to doing."

"Do you mean grand mom can't do anything? OH NO!"

Mom goes on, "well, it is too soon to tell how she will be, but right now, she is pretty sick."

"Can I visit grand mom?"

"Not yet, honey."

This news has really made me sad.
The days are not as bright, even the
sunny ones.

"Today, I am getting to facetime with Grand mom! I am seeing her for the first time! I am so excited! I will wear her favorite dress for the event."

My mom came home with good news, that Grandmom is moving to a different sort of place, a "rehabilitation" hospital.

There, many specialists will help her get better enough to come *back* home to us. The good part, is that I get to visit her in this hospital!

Today is the first day I see my grandmom face to face! I will put on my favorite outfit today. I can't wait to see her!

13

Grand mom looks better today! When I get there, there is a person with her, who is showing her how to get in and out of bed by herself.

14

Each day I go to see grandmom, she looks better and better!

15

I never knew about this type of hospital! There are many different kinds of health care workers, doctors, nurse practitioners, nurses, physical therapists, speech therapists, occupational therapists, nursing assistants, dietitians, and dietary workers. All these people helped my grand mom get much better!

16

The doctors and nurse practitioners coordinate the care for grandmom so that activities and treatments are scheduled each day to help grandmom get strong in many ways.

Nurses-Develop a plan of care for patients to he them get better and or feel better. Nurses work with many members of the health care team to achieve this.

Because of the time they spend with patients, they often have the best understanding of how a person may be coping with a disease or illness, and are key to helping people cope.

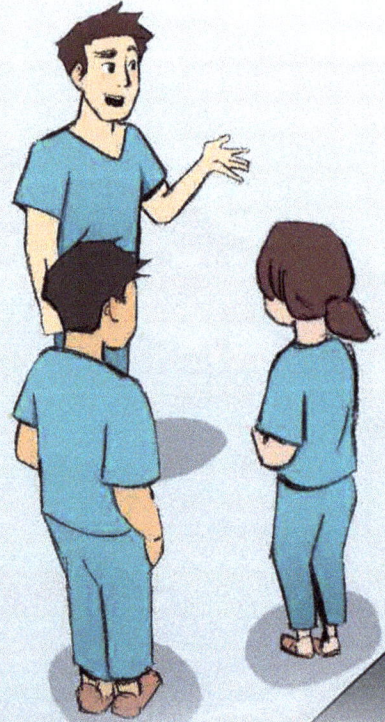

They do lots of stuff like checking to see how some is doing physically, informing the right people about early status changes and documenting the effectiveness of the plan to care for someone.

18

Physical therapists help people recover from illness or injury by helping them improve the way they move and manage pain. Physical therapists are experts at understanding how the body moves and different ways to strengthen and restore function.

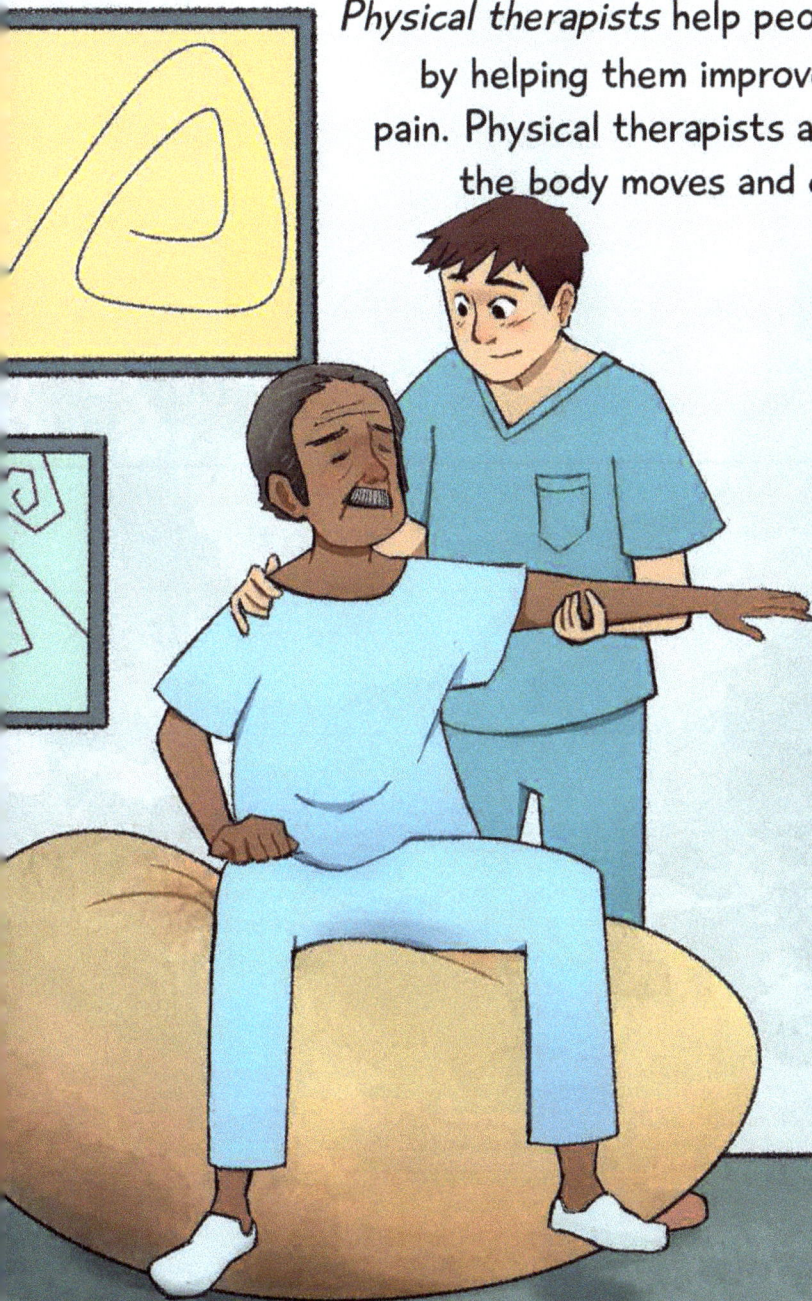

Speech therapists are people who help people who develop problems with speech, language, voice problems recover from those problems, or develop new ways of communicating. They are officially referred to as speech-language pathologists.

20

Occupational therapists develop a plan to help people who are recovering from illness or injury get better by working with them to regain independence.

They develop an individualized plan to help people with their activities of daily living (ADLs). They adapt the teaching plan to the person's environment and work with any limitations that might require new ways of doing things.

21

Nursing assistants help nurses with the many tasks involved with caring for people like baths and walking around, eating, *checking* vital signs, and so much more.

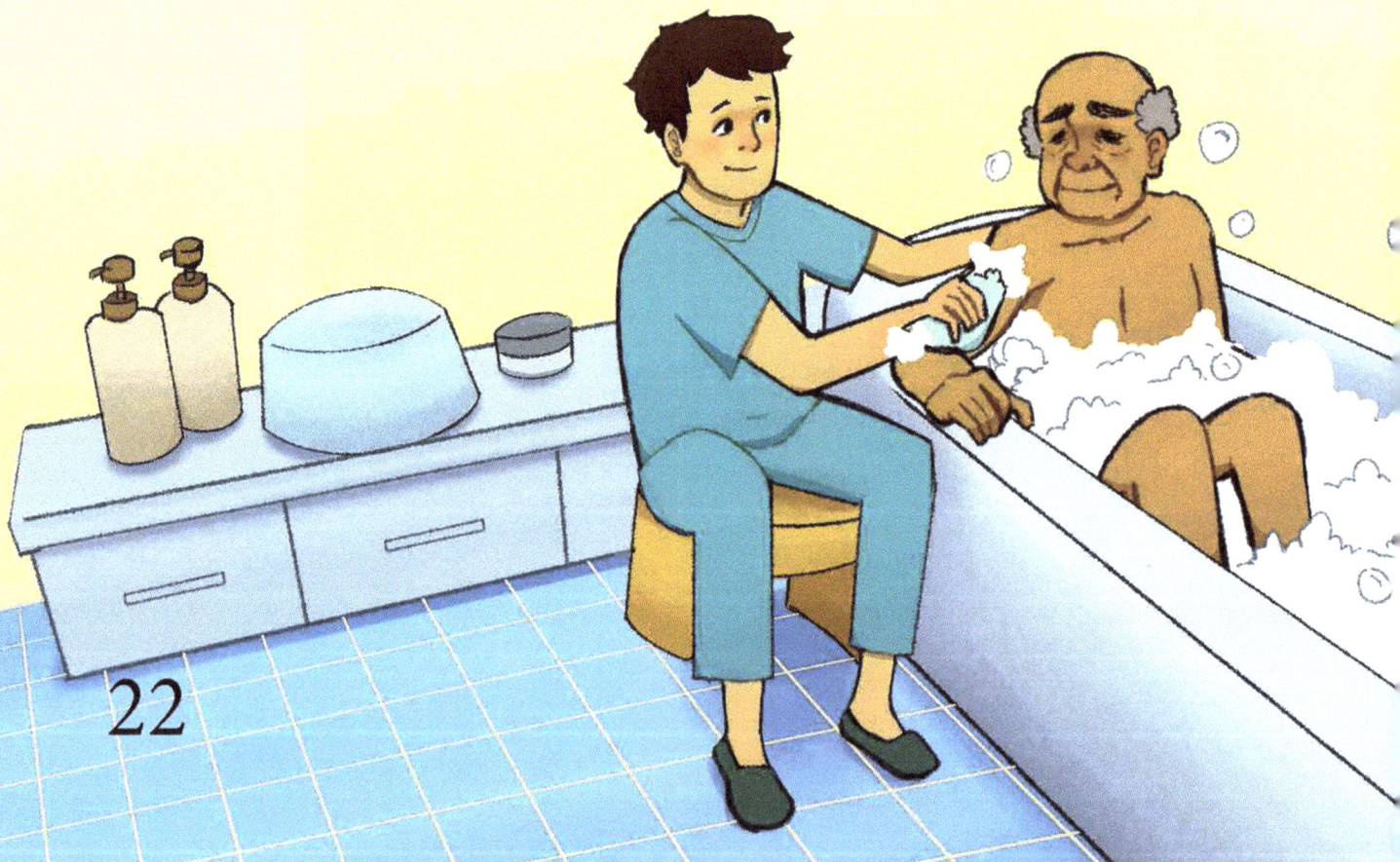

Dietitians   Dietary workers   Housekeeping

23

Seeing all these people work together in the rehab hospital is like watching an orchestra, but instead of playing instruments, everyone is doing their part to get people back to their living space as independently as possible.

About the Author

Glossary of Helpful Words

Stroke
A serious problem in the brain that can make it hard to talk, walk, or move.

Rehabilitation
Special care that helps someone get better after being sick or hurt.

Hospital
A place where people go to get help when they are very sick or hurt.

Physical Therapist
A person who helps people move better and feel less pain.

Speech Therapist
A person who helps people talk and understand words better.

Occupational Therapist
A person who helps people do everyday things like getting dressed or cooking.

Nurse
A person who takes care of patients and helps them feel better.

Health Care Team
A group of people who work together to help someone get better.

Vital Signs
Important body signs like heartbeat, and temperature that show how healthy someone is.

Nursing Assistant
A helper who supports nurses with things like bathing and walking.

Dietitian
A person who helps people choose healthy foods to stay strong.

Dietary Worker
Someone who prepares and serves food in hospitals.

Housekeeping
People who keep the hospital clean and safe.

Facetime
A video call where you can see and talk to someone on a screen.

Droopy
Hanging down or looking weak — like when part of someone's face doesn't move well.

Independence
Being able to do things by yourself.

Plan of Care
A special plan made to help someone heal and feel better.

Dr. Clary-Muronda

has been a nurse for over 30 years. A Philadelphia native from the West Oak Lane section of the city, she attended Albert Einstein School of Nursing, and continued on with her education over the years and later earned a PhD in Nursing Science from the Medical University of South Carolina in Charleston. She enjoys teaching undergraduate nursing students, mentoring young students who aspire to work in the health professions, and inspiring nurses to be the best they can be.

Also by the same authour Valerie Clary Muronda

Why
Keisha
wanted to become
a Nurse

Why Keisha wanted to Become a
Nurse

By Valerie Clary Muronda

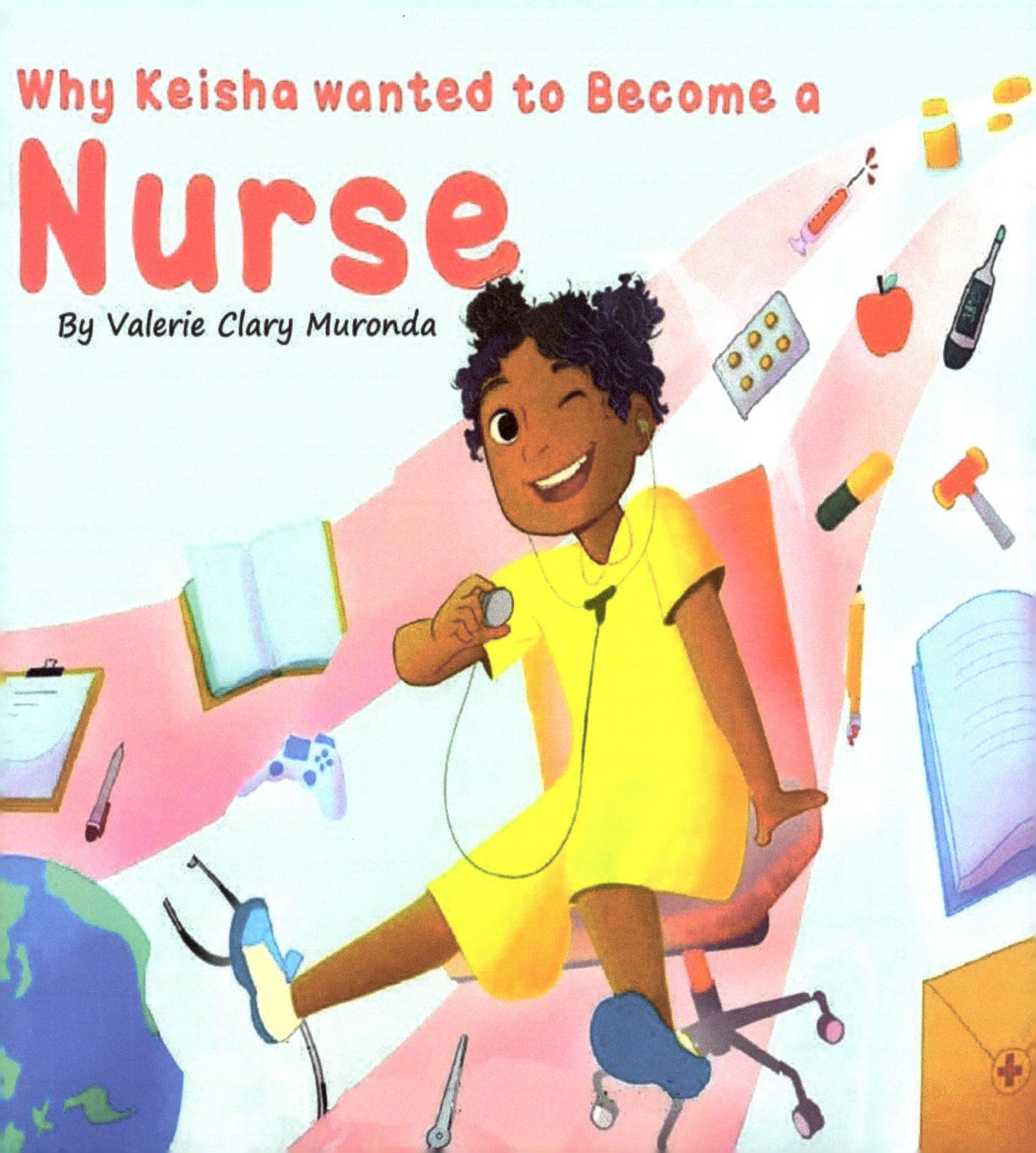

Sometimes Keisha has trouble breathing because of asthma. So she has been to the hospital many times. She likes the way the nurses help her get better.

Now Keisha is thinking about becoming a nurse when she grows up so, she can help kids get better too!

Nurses Rock!

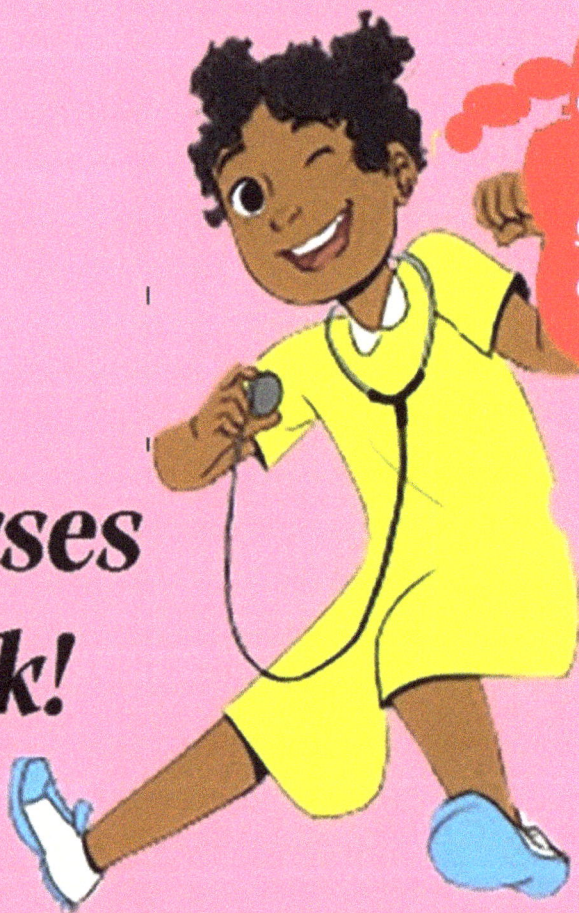

www.vclarymuronda.com

www.ingramcontent.com/pod-product-compliance
Lightning Source LLC
Chambersburg PA
CBHW042348030426

42335CB00031B/3496